THE LIMESTONE HERITAGE

Siġġiewi

VINCENT ZAMMIT

PHOTOGRAPHY
DANIEL CILIA

HERITAGE BOOKS
2004

HOW TO GET TO THE LIMESTONE HERITAGE

By *bus:*
No. 89 from Valletta bus terminus stopping at Siġġiewi main square. Continue on foot going through Saura street and Mgr M. Azzopardi street, following the road signs.

By *car:*
Main roads leading to Żebbuġ and Siġġiewi. On leaving the Żebbuġ by-pass follow the road signs.

The Limestone Heritage
Mgr. M. Azzopardi Street, Siġġiewi QRM 14
Tel: 2146 4931
Fax: 2146 0153
www.limestoneheritage.org

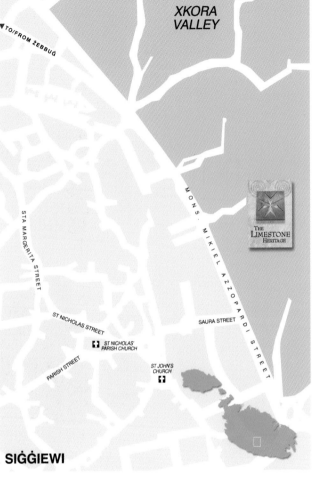

Insight Heritage Guides Series No: 7
General Editor: Louis J. Scerri

Published by Heritage Books, a subsidiary of
Midsea Books Ltd, Carmelites Street,
Sta Venera HMR 11, Malta

*Insight Heritage Guides is a series of books intended
to give an insight into aspects and sites of Malta's
rich heritage, culture and traditions.*

Produced by Mizzi Design & Graphic Services
Printed by Gutenberg Press

Copyright © Heritage Books
Photography © Daniel Cilia

First published 2004

ISBN: 99932-39-96-8

OUR HISTORY WRITTEN IN STONE

One of the few, if not the only, natural resource of the Maltese islands is its stone. Looking around the built-up areas, one notices an impressive array of decorated stonework, ranging from prehistoric to modern times. The stone has been and still is an important source material for the building trade, and for those artisans who treat it with love to create impressive sculptures. The journey we are going to take will enable us to understand and appreciate better this important distinctive heritage.

The story of Malta starts with the tectonic plate movements that occurred in the Mediterranean area. The islands we call Malta had not yet been formed. The sedimentary stone of these islands in fact indicates how these were formed and how they took their particular configuration. Changes in temperatures and other climatic conditions helped to form the islands' geological features with its rugged coastline, cliffs, caves, and a

limestone heritage that has been a natural resource for the building industry since prehistoric times.

Limestone started being used in earnest by the first people to inhabit the islands and who lived either in natural limestone caves or in huts built

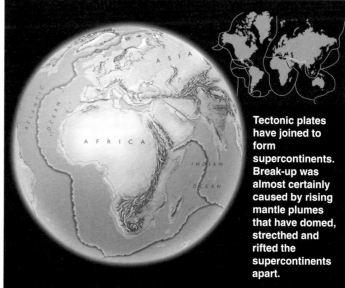

About 5 million years ago, a rise in the sea level allowed Atlantic waters to flood the Mediterranean basin

Tectonic plates have joined to form supercontinents. Break-up was almost certainly caused by rising mantle plumes that have domed, stretched and rifted the supercontinents apart.

Prehistoric builders were skilful engineers and master craftsmen

Opposite:
Siġġiewi, a village of farmers and mason workers

understanding of Maltese architectural heritage. During these millennia, the first big structures were erected which have even been declared as the *oldest free-standing buildings in the world* and acknowledged as UNESCO World Heritage Sites. They amply demonstrate the level of sophistication of these first Maltese builders.

Limestone is easily quarried. Even though sophisticated tools were not available, it was still possible to quarry the stone and use it according to the needs of the community. Unfortunately, no prehistoric quarries have been discovered, in spite of the extensive quarrying activity that took place during the time. Indeed all the prehistoric temples were constructed with stone quarried some distance away, a detail which needs to be pointed out. With their unsophisticated tools the process must have been difficult and time consuming, but at the same time the prehistoric Maltese builders

in the open. Ghar Dalam, Birzebbugia, is reputed to have been one of the earliest caves to be used for habitation. On the other hand, the open village of Skorba, also utilized by these first colonizers, consists of built huts. These settlers made use of the stone available in the area, and learned how to utilize this natural source.

The prehistoric era is very important for the study and

Top: **A model in stone of a prehistoric temple**

Right: **Ġgantija temple in Gozo**

The remains of ancient Roman quarries in the Dingli-Rabat area

understood the nature of the local limestone, and thus quarried it without any major problems.

The first local quarries that have been identified date to Roman times. The Romans carried out extensive building activity in and around the capital city of Melita (present Mdina and Rabat). A number of Roman quarries can be seen in the surrounding areas of this urban centre.

Close to these quarries, the famous *cart-ruts* can be observed. Although these cart-ruts have never been satisfactorily dated, some scholars believe that they belong to Roman times. However, there seems to be a consensus that the cart-ruts were used to carry heavy loads and probably stone as well. The building that occurred during Roman times all over the islands could been another boom period for stone quarrying. Although the urban centre lay in the centre of each island, various other buildings were erected all over the islands, either for defensive purposes or as extensive farmsteads.

The building trade seems to have lost some of its vigour in the following centuries. It is very difficult to imagine the building of anything other than small huts and houses. The period was not a peaceful one and this did not allow the Maltese to settle down peacefully and think of improving their environment. The difficult economic situation and the continual attacks by pirates led the Maltese to refrain from building and decorating elaborate houses.

The middle ages brought about a slight change. Some fortifications were built, or at least improved. The Grand Harbour's *castrum maris*, which was to become Fort St Angelo, Mdina, and the *castello* in Gozo were the only large-scale building structures undertaken during these times. The local authorities regularly complained of the lack of necessary funds for the upkeep of these defences.

The change came with the arrival of the Order of St John in 1530. The Maltese islands then had a population of about 18,000. With the security

The imposing Fort St Angelo in the Grand Harbour

afforded by this military Order, better employment opportunities, adequate defence, and health services, the population continued to increase. This increase led to the demand for more housing and whole cities and villages were established.

Extensive defence projects also increased the demand for the local natural resource, namely limestone. The limestone was soft enough to be easily worked, while being strong enough to withstand cannon shot. The first defences were erected around the Grand Harbour, but after the successful outcome of the Great Siege in 1565, there began the greatest building period ever in Maltese history with the building of the city of Valletta.

Valletta needed to be provided first with a line of fortifications, which were carved out of the living rock, with further stone courses being added on top to provide a unified line of defence. The fortifications completed, in the second half of the sixteenth century the builders turned their attention to the palaces, churches, and the other buildings that befit a city. The style of architecture was not one with elaborate decorations. However, at the turn of the seventeenth century various architectural details started being introduced. Church façades did not remain as plain as those of the previous century. The elaborate decorations which became the norm were the forerunners of the baroque period. The churches and palaces that had been erected a century earlier started to be redecorated, or even rebuilt. Fortifications continued to be built throughout the years that the knights were in Malta. All of this building activity created considerable economic activity.

Top: An aerial view of Valletta and its unique line of fortifications and bastions

Above: St Nicholas old parish church in Siġġiewi, attributed to Tumas Dingli (1591-1666) now in ruins

Quarrying was at its height during this time. Many of these quarries have either been identified or else are still extant even if not really all that well known. In order to hurry up the building of the fortifications, the knights sought to quarry the stone from where they were going to make their ditches. This means that all the ditches of the fortifications of this period were actually the quarries. Another quarry was planned for the Marsamxett side of Valletta. The idea was to excavate this place to below sea-level and, by breaching the rock face, a safe inland harbour within the fortifications of the new city would be created. The quarry was excavated to a certain level, but the project was then abandoned as the stone was not of the required standard.

Quarrying during the time of the knights was carried out in a manner that survived until the early twentieth century. Modern technology started to replace the older and almost primitive tools following the Second World War. New machinery and tools were imported to replace the manual tools of former years.

For many centuries quarrying in Malta had been carried without any drastic change. The work was done manually in shallow pits with the quarried stone being taken to the actual building site on mule-drawn cart. Whenever possible, the building site itself was used as a quarry but, as the need for more stone was felt, a different place, not always nearby, would be chosen

An important trade with close connection to the building trade was the blacksmith's. Tools needed to be made, updated, and checked by the blacksmith on a regular basis. His was a very important aspect of this trade, as without such tools, quarrying could not take place. Blacksmiths became so proud of their job that they even included their own special mark on the tools they made. The pride and personal satisfaction of these tradesmen benefited the trade in general.

Debris was carried away in wicker baskets. The craftsmen who made

Opposite: **Although machanized, work in the quarry still remains labour intensive**

Below: **A blacksmith**

Bottom: **The Limestone Heritage Museum which houses the original tools used in the construction industry**

A modern quarry

A converted farmhouse

Each *vjegg* consisted of a particular number of stone slabs, about 30 on each cartload. Payment was controlled by the number of stones received on the building site.

The quarrying industry started to become more mechanized following the Second World War, with machines becoming easier to install and therefore more popular. The means of transportation also improved, making work easier and more efficient. Although work in quarries was still considered to be quite hard, it now became more humane. The amount of post-war reconstruction going on in Malta increased considerably the number of quarry workers and helped in the introduction of modern technology.

This period also saw the introduction of new building material and methods. Concrete started being used more, with bricks becoming more common in the building trade. This may have led many to believe that the old style of building would eventually die out but, because of the modern

these baskets were not considered as important as the blacksmiths, although their contribution was necessary for the better management of the hard work at the quarry. There was also the need for transport facilities to move around the quarried stone, which was provided by mule-drawn carts. This was the origin of the famous 'load', or *vjeġġ* in Maltese.

interest in restoring old town houses and farmhouses, the old styles, building techniques, and even technical words are actually coming back into use.

Quarry workers needed to be young and strong. They had to start early in the morning, as working at the bottom of a quarry would be impossible in the midday heat. Moreover, after their hard work at the quarry, many of these workmen would return to tend their small family fields.

Quarries eventually end up being unproductive. Some of these old and disused quarries are simply abandoned, while others have been turned into excellent orchards, protected from the winds, with adequate supply of water from cisterns. In 'The Limestone Heritage', part of the quarry has been changed into a field, still tended and taken care of by the family.

THE GEOLOGICAL STRATA OF THE MALTESE ISLANDS

The rocks of the Maltese islands are all sedimentary, which means that their history, in geological terms, is quite recent. They were formed in relative shallow marine environment, and this also explains why the presence of numerous fossils of sea creatures. The rocks that form the islands all originated from sheets of unconsolidated sediments which were laid at the bottom of the seas, between 24 million years to about 6 million years ago. As a result of plate tectonics, the land surfaced, and thus provided the necessary temperatures and conditions for the sediments to harden.

The Maltese islands, like all the other central Mediterranean islands, are made up of sedimentary rocks, mainly limestone. Four main layers of rocks have been identified in the Maltese geological formation, with a small, almost insignificant fifth one included in between. Starting from the lower level, there is the LOWER CORALLINE LIMESTONE. This has been extensively used for kerbstones, doorsills, stairs, and even monuments. One of the most prominent monuments is the War Memorial just outside Valletta which was made entirely from this stone. This stone was also employed to make the stairs of the main palaces of the knights. All varieties of this hard stone are used in the building industry and in road-surfacing.

The next layer consists of GLOBIGERINA LIMESTONE and is a softer and more yellowish stratum. The word 'Globigerina' derives from the common name for a type of microscopically small fossil shell which is abundant in this limestone. This level is subdivided into three others which are usually referred to as the Lower, Middle, and Upper divisions of Globigerina Limestone. The Lower layer provides the most important of all this stone. Known as *franka*, this is extensively used for building and ornamental purposes. The Middle layer is made up of white-weathering grey stone, and is usually of no use as a building stone. However, a stone known as 'chert' can be found inside it. This is the only hard silicone rock found in Malta and was used extensively in prehistoric Malta. The Upper layer of Globigerina Limestone is subdivided into yet another three other sections. Since ancient times it has been known that certain sections of this layer is quite heat resistant and has been used to build ovens, and to make the small Maltese ovens, known as *kniener* (singular *kenur*).

The third layer is that of BLUE CLAY. Although Maltese clay is not of the best quality, yet it forms an integral part of our landscape. As it actually forms part of an important impervious base, it makes natural water springs possible. As the layer above the Blue Clay, namely the Upper Coralline Limestone, is a water-bearer one, it offers the possibility to have such springs, as well as the important aquifers. It has also helped in the formation of the characteristic valleys of the islands.

Just above this layer, there is the small and almost insignificant one known as GREENSAND which is not to be found all over the island. It is usually only encountered in thin layers. In most cases, Greensand forms the base of the hard and resistant Upper Coralline Limestone.

The topmost Maltese stratum, known as UPPER CORALLINE

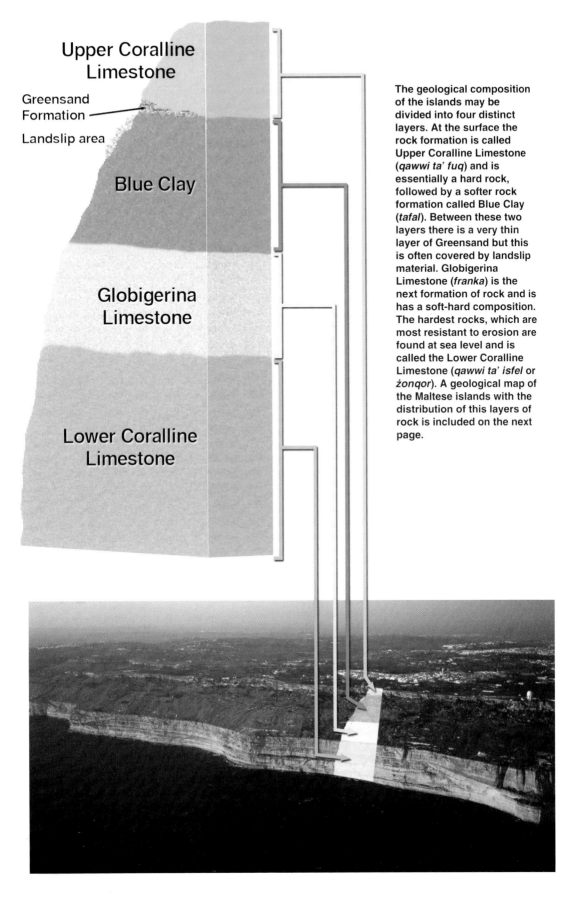

Upper Coralline Limestone

Greensand Formation

Landslip area

Blue Clay

Globigerina Limestone

Lower Coralline Limestone

The geological composition of the islands may be divided into four distinct layers. At the surface the rock formation is called Upper Coralline Limestone (*qawwi ta' fuq*) and is essentially a hard rock, followed by a softer rock formation called Blue Clay (*tafal*). Between these two layers there is a very thin layer of Greensand but this is often covered by landslip material. Globigerina Limestone (*franka*) is the next formation of rock and is has a soft-hard composition. The hardest rocks, which are most resistant to erosion are found at sea level and is called the Lower Coralline Limestone (*qawwi ta' isfel* or *żonqor*). A geological map of the Maltese islands with the distribution of this layers of rock is included on the next page.

LIMESTONE, is the youngest of all. Its most important function is that of providing aquifers for the islands' vital water resources. Moreover, it has also been used for building purposes, just like Lower Coralline Limestone. The soft pure limestone of the lower division of this formation has been extensively used to make lime for the building industry.

STONE QUARRYING is the extraction of rock to be used for different needs. This is not simply the cutting, quarrying, and working of stone, but there needs to be an understanding of the qualities and capabilities of the actual stone. The quarrying carried out by the Maltese is the result of experience gained through the centuries of activity on the islands. This process is carried out all over the world, but the method is influenced by the particular type of rock found in the area and by the use envisaged. The stone quarried varies, as not all stone is good for building purposes. Experts point out different uses for the different types of stone that originate from particular areas. There are also different methods to extract stone, depending on the different types and uses. Explosives are usually used to shatter the rock which is intended for industrial purposes. Such broken rock is used to make cement, to mix with concrete, or to form roadbeds. Sometimes the rocks are shattered into even smaller pieces, for specific needs. Harder rocks are also quarried by blasting. Other methods are used for softer stone and depend on whether the stone is needed for decorative purposes.

Quarrying can be carried out wherever the Globigerina Limestone layer is to be found close to the surface. Most the quarries in Malta lie in the Mqabba-Qrendi-Siggiewi and the Naxxar-Msierah areas. Not all Globigerina stone is good for building purposes, with the best being known as *franka*. When freshly cut, this stone is very soft and easily dressed and worked but in time, it hardens up and becomes strong and weather-resistant.

Lower Coralline Limestone is mostly quarried along the Great Fault near the Mosta-Naxxar-Gharghur area. Upper Coralline Limestone, a similar rock formation to the one previously mentioned, is mostly found in the Mellieha-Marfa ridge area.

The quarry's machinery and tools depend on the type of stone it has, as can be seen in the various quarries that are still functioning to this very day. Stones are cut and sent to building sites ready to be used. Only slight finishing touches are needed. Harder stone is crushed and prepared according to the needs of industry.

N

A geological map of the Maltese islands

Upper Coralline Limestone

Blue Clay

Globigerina Limestone

Lower Coralline Limestone

THE LIMESTONE HERITAGE

Since quarries are dangerous places they are generally not open to the public. Fortunately, a disused quarry at Siggiewi has been turned into one of the most interesting and educational heritage trails in Malta and a fascinating experience for tourists and locals alike. Visitors enjoy the opportunity of entering a real quarry, seeing how stone was cut through the ages, and how the trade is still practised to this very day. The Limestone Heritage is a family-run attraction. The husband-and-wife team has its family origins in the building industry. The husband is the son of the former president of the Quarry Owners Association of Malta, whose family ran and operated quarries for generations. The wife is the daughter of a leading building contractor, whose grandfather and ancestors have been involved in the building trade for years.

Quarrying started by having the workmen marking the area where they were going to work. Trenches, about one metre deep and two metres wide, would be laid out. This was the very start of the long process of producing stone. The debris from these initial trenching works would be carried away in wicker baskets and kept aside, as it might be used elsewhere. The next step was to start actually cutting the stone, by separating it from the bedrock. Slots were dug at 50 cm intervals and then metal plates inserted in these slots. Other heavy metal wedges were used to break off the stone. Sometimes wooden wedges were also used. These were soaked in water which would lead the wood to expand during the night and break the stone off.

Limestone Heritage, converting a disused quarry

Carrying the debris in wicker baskets

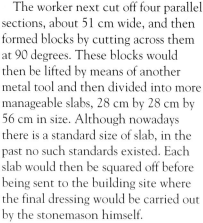

The worker next cut off four parallel sections, about 51 cm wide, and then formed blocks by cutting across them at 90 degrees. These blocks would then be lifted by means of another metal tool and then divided into more manageable slabs, 28 cm by 28 cm by 56 cm in size. Although nowadays there is a standard size of slab, in the past no such standards existed. Each slab would then be squared off before being sent to the building site where the final dressing would be carried out by the stonemason himself.

Although there were no set sizes, it was customary to speak of a particular measurement, when stone-cutting was being carried out. The *xiber* was the accepted unit which was equivalent to the outstretched hand span of a worker, about 26 cm.

The stone slabs were then loaded onto mule-drawn carts to be carried to the building site, often some miles away. The cart was loaded with a *vjegg*, which was about 20 slabs of today's standard-sized slabs. The load depended on the size of the actual

Separating the stone from the bedrock (*above and opposite*), digging of slots and trenches (*right*) and use of heavy metal wedges (*below*)

Frans Baldacchino

Born into a family of quarry owners and builders in 1921, Frans Baldacchino started working within the industry in the early 1950s. At that time, the war reconstruction programme was still going on. Moreover in the late 1930s and early 1960s the tourism industry was being given the necessary official backing to prepare for the expected influx of tourists to be brought to Malta. This led to major buildings projects in new hotels, restaurants, and other tourist facilities.

During this time, Frans Baldacchino lived through the important changes that were taking place within the quarrying industry. Manual jobs were being made redundant by the introduction of new machinery, which increased efficiency, decreased hardship, and improved the working environment. This led to better working conditions, and increased profit for all concerned thanks to a better organization within the industry. Frans Baldacchino, as president of the Quarries Owners Association during the 1970s and 1980s, was at the heart of these activities and changes. His contribution towards the Association has been acknowledged as enormous.

Frans Baldacchino died on 13 January 1999 at the age of 78.

Preparing the stone slabs into standard sizes

The circular saw used for cutting the stone from the bedrock, easing the manual work at the quarry

stone slabs, as sometimes a stonemason needed different-sized slabs for a particular section of the building.

The advent of mechanized tools from Italy changed the industry. These early machines were heavy and noisy; nevertheless, they were more efficient and precise thus increasing the production of building stone. The giant circular saw could be easily moved along the quarry floor over rail tracks. The saw first cut long parallel vertical gutters to determine the width of the slabs; then the same process was repeated at an angle of 90 degrees to determine their length; finally, the saw was turned horizontally to separate the ready-to-use slabs from the quarry-face.

This new mechanized process marked the walls of the quarry with vertical and horizontal grooves from the saw.

The modernization of stone transportation improved work further in the quarry. The diesel-powered, 1939 Dodge lorry replaced the mule-drawn cart as it could carry larger loads. The lorry was adjusted to have its three sides that could open to facilitate the loading and unloading of stone.

The lorry soon became part of the Maltese road cult. It was not only a means to transport stone but became an extension of the driver's personality and an expression of his character. Owners started decorating their trucks in the same style and fashion as they had done with the carts. The same trimmings soon started to appear on the bonnet and cabin, while the interior of the cabin was decorated in the driver's personal taste with curtain fringes, lucky charms, and pictures. In the latter case, these changed from time to time, starting with holy pictures, to pop and film stars, to pin-up girls. The sense of belonging was so great with these lorries that their owners gave them names that they flashed on the side of the bonnet. Naming these trucks was like naming a child and in this case as well the choice was varied from names

of saints to pop and film stars, marking the passage of time.

During the same time, a particular hue of green, reminiscent of Georgian green, seems to have been unintentionally chosen to paint these lorries, and it became a common feature on Maltese roads and building sites.

On one of the quarry faces there is a cross-section of a bell-shaped well. This shape of well is very popular in Malta and similar ones have been found in Roman sites. With Malta's dry climate, a regular water supply is vital. Building regulations during the construction of Valletta stipulated that every house had to have its own well. Moreover, the stone excavated from excavating such wells was later used in the building of the house or palace where the wells were situated. This ingenious method saved both time and money for those building houses and palaces in Valletta.

Various geological strata can be noticed in the same quarry face. These geological features are discussed in detail on pages 12, 13, and 14.

A further drive to modernize quarrying was experienced in the sixties when the demand for building and stone was at its height, also owing to the vast expansion of the tourist industry that had introduced Malta as a holiday destination. Modern technology meant faster and more efficient machines. One such improvement was the electric conveyer belt which reduced the hardship of loading the stone on trucks.

This technology also introduced equipment which helped workers quarry deeper into solid rock. The quarry face is a witness of this passage

The lorry, with its distinctive marks of the owner, replaced the cart

The bell-shaped well

Opposite: **The orchard is a frequent end of a disused quarry**

Fossils, a frequent find in quarries

of time and the different technologies used – a smooth quarry face denotes the use of modern machinery. However, work was not always easy. An underground stream or cavity, looking like small caves complete with stalagmites and stalactites, created numerous problems when encountered in a quarry. Coming across such cavities made it necessary to change the saw blade, since the limestone in these cavities would have hardened owing to calcification. It also resulted in loss of work time since it was necessary to make new calculations to work around such natural abnormalities. At the end of this quarry there is an example of an outcrop which has been retained in its original state.

Many old and disused quarries ended up being turned into small orchards or fields. A section of the quarry being visited has been turned into such an orchard. Besides noticing the various fruit trees, there is also a clear indication of the stratigraphy of the soil mixed with rubble stones. This mixture creates a very good draining system.

At this point there is also another important detail one should notice. The quarry floor is littered with a number of fossils, which means that millions of years ago, before the creation of the present configurations of the Mediterranean coastline, this used to be the bed of an ancient sea. Fossils are common in Maltese rock strata, and they are a clear indication of the origin of Malta's stone. All kinds of fossils are to be met with in Malta and Gozo.

The visit to the quarry will continue with a look at the various building techniques utilized over

Limestone Heritage visitors' centre and museum

the centuries. As has already been indicated, the contemporary fashion of restoring old farmhouses and town houses to their original grand style has helped to bring back old building techniques have become important and relevant once more.

Maltese fields are usually characterized by rubble walls which separate the various fields and help retain the scarce soil. Few structures, except for some farmhouses, are to be noticed in the countryside. The smaller structures known as *giren* (singular *girna*) are

usually built of rubble stones. This small corbelled structure has always been meant to be a small and functioning field hut. The nearby building shows the various interesting architectural features of the Maltese *razzett* and of Maltese townhouses. Each and every part of the building is detailed and fashioned according to the needs of the owner. At the same time one can clearly see the great adaptability of Maltese worker, who uses his natural resource to create a better environment for himself and his family.

TYPICAL MALTESE BUILDING FEATURES

The rubble walls and the *girna*: the Maltese countryside cannot be said to be extensive, and neither can it be called impressive. Yet, it has got its own characteristics which marks it out as a different component of the larger and wider Mediterranean type of countryside. The main characteristics are encountered even in Malta, but then, as in many other things, the situation changes and it obtains its particular character.

Two characteristic items of the islands are the rubble walls (*ħitan tas-sejjieħ*) and the field huts (*giren*). Rubble walls are found around every field, generally separating fields into small units, thus making it possible for different family members to know which part belongs to them. The walls are usually found in terraced fields that are so characteristic of the countryside. These walls have also another important function, in that they build and create fields on the gentle slopes of the Maltese countryside and help to retain the precious little soil there is.

These rubble walls are made with various-sized stones found in the fields or brought from areas nearby. There must have been a lot of moving of such stones from one place to another to make these walls. It is necessary to make sure that there is always some room between the stones to make it possible for water to percolate easily. At the same time, these walls help to retain the little soil in the fields, while making it possible for the lower fields to share the available water.

Another characteristic of Maltese fields are the *giren*. Although today there are also modern structures, the most interesting and most spectacular remain the older ones. *Giren* are built from loose stones found in the fields, just like the ones used to build the boundary walls. These huts are constructed in various shapes and sizes, but they were never meant to

The intricate inside of the corbelled dome-like roof of the *girna*

replace the other Maltese countryside feature, the *razzett* or farmhouse.

Most *giren* are one-roomed and were erected to shelter the farmers and the other field workers from inclement weather and to store their tools for the day. Some *giren* are built round, while others have an oval shape. Some, which tend to be slightly bigger, have been a staircase or even a buttressing ramp on the outside. A few *giren* are rectangular in shape. Each *girna* tends to have one door, with the keystone above the entrance being the largest of the lot. The other stones are small in size, of differing sizes, and carefully placed on each other. Although such a structure may appear crude, it demonstrates the intelligent preparations needed to plan and build them.

The corbelled dome-like roof is another interesting feature of these structures. This technique is reminiscent of the method used to roof the temples of prehistoric Malta.

In spite of the impression of flimsiness, the *girna* is quite a sturdy

structure, and only starts to crumble when it is abandoned. The structure shows the great ability of the builder who knows the type and size of the stones that need to be used, and how to fit them together. The builder usually is quite well-versed in his trade, and therefore knows that one very important feature of the structure is the filling he puts in between the outer and inner walls. In fact, it is when this filling starts to crack up and is not taken care off immediately, that the *girna* starts to collapse.

The *giren* are still a very characteristic feature of the Maltese countryside and it would be a real shame if they were to be allowed to fall into a heap of rubble. The same thing can be said about the rubble walls. These two manifestations of vernacular architecture have not only become part of the national character, but they also provide evidence of the way that the Maltese have always managed to make use of the most important natural resource of their islands – the limestone.

The *girna*, a typical rural structure found in the countryside

The Maltese farmhouse, or *razzett* as it is known in Maltese, is an outstanding and unique example of rural architecture in the Maltese islands. It has often been described as 'architecture without architects' because it owes its development to the ingenious farmer and mason who sought to develop an abode fit for his needs and those of the environment that surrounds him. The following are some examples of traditional building terms.

Il-maxtura, manger, a trough used for feeding animals

Imramma, building debris used to fill-in between thick walls

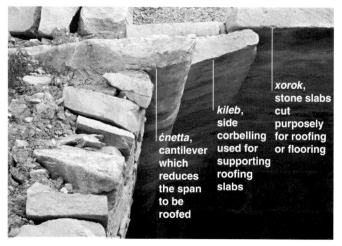

ċnetta, cantilever which reduces the span to be roofed

kileb, side corbelling used for supporting roofing slabs

xorok, stone slabs cut purposely for roofing or flooring

 *Ħnejjiet* (*ħnejja*) or *arkata*, types of arches to support the roof

 *Travu ta' l-injam*, wooden beam, another way to support the roof

 *Il-moħba*, a small niche next to the roof, usually behind a beam to hide away valuables

 *Tarjola*, a pulley, using the holes at the end of a galley's mast used as a beam

 *In-niċċa*, a devotional shrine or niche, smaller ones were used to keep oil lamps

 *Armarju tal-ħajt*, a wall cupboard

 *Ċumnija*, chimneys used to extract fumes from the kitchen stove

 *Barumbara*, a pigeon loft

 *Miżieb*, a water spout, carrying water from the roof to clear the wall

 *Kanal*, a stone gutter used for irrigation

Ħawt, a water trough

 *Spurgatur*, an overfow

 *Marbat*, a stone ring to which animals are tied

 *Herża* or *ħorża*, a well head or curb

 *Plier*, two side pillars with a flat monolith on top bridging the gap from which the pulley hangs

Rewwieħa, a small opening used for ventilation

Garigor, a stone spiral staircase

Newwieħa, used to let the domestic cat in and out of the house at will

Il-harrieġa, a projected stone shelf, is a distictive feature in rural buildings and inserted, often in pairs, usually near the main entrance. It served as a secure place where to place a lamp, out of reach from children, or more romanti- cally an ideal place to display potted flowers.

A special feature found in the farm house reconstructed at the Limestone Heritage, is the kitchen. The *fuklar*, a stone cooker (*below*) and the *forn*, oven fired by twigs, thistles, and bushes. The *kenur*, was a hand-carved stove (*right*) used for slow cooking

TRADITIONAL TOOLS USED
IN THE BUILDING INDUSTRY

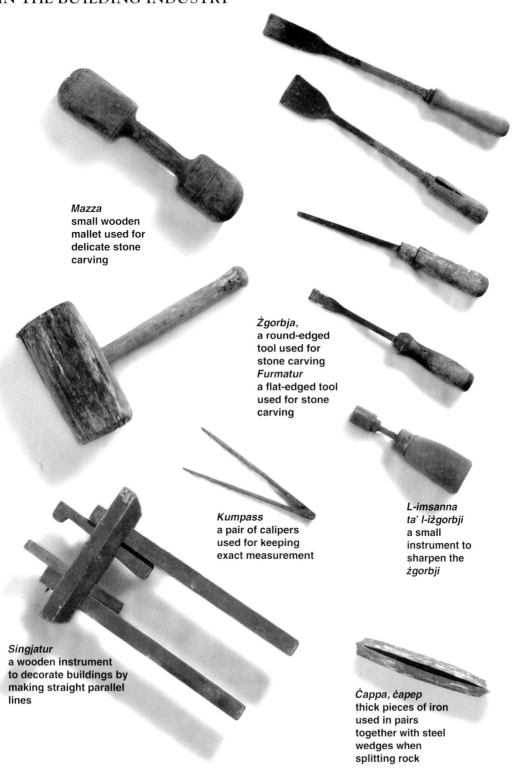

Mazza
small wooden
mallet used for
delicate stone
carving

Żgorbja,
a round-edged
tool used for
stone carving
Furmatur
a flat-edged tool
used for stone
carving

Kumpass
a pair of calipers
used for keeping
exact measurement

**L-imsanna
ta' l-iżgorbji**
a small
instrument to
sharpen the
żgorbji

Singjatur
a wooden instrument
to decorate buildings by
making straight parallel
lines

Ċappa, ċapep
thick pieces of iron
used in pairs
together with steel
wedges when
splitting rock

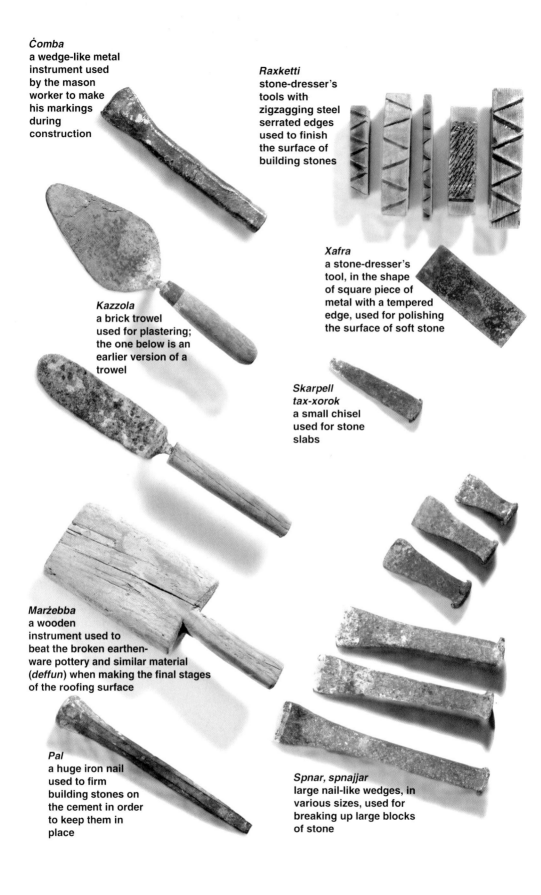

Ċomba
a wedge-like metal instrument used by the mason worker to make his markings during construction

Raxketti
stone-dresser's tools with zigzagging steel serrated edges used to finish the surface of building stones

Xafra
a stone-dresser's tool, in the shape of square piece of metal with a tempered edge, used for polishing the surface of soft stone

Kazzola
a brick trowel used for plastering; the one below is an earlier version of a trowel

Skarpell tax-xorok
a small chisel used for stone slabs

Marżebba
a wooden instrument used to beat the broken earthenware pottery and similar material (*deffun*) when making the final stages of the roofing surface

Pal
a huge iron nail used to firm building stones on the cement in order to keep them in place

Spnar, spnajjar
large nail-like wedges, in various sizes, used for breaking up large blocks of stone

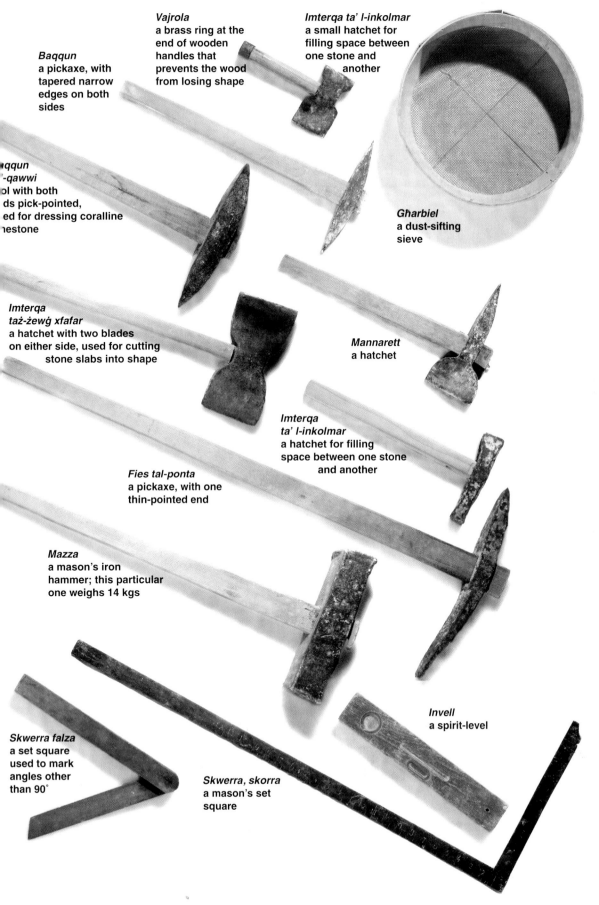

Baqqun
a pickaxe, with tapered narrow edges on both sides

Vajrola
a brass ring at the end of wooden handles that prevents the wood from losing shape

Imterqa ta' l-inkolmar
a small hatchet for filling space between one stone and another

**...qqun
...-qawwi**
...l with both
...ds pick-pointed,
...ed for dressing coralline
...estone

Għarbiel
a dust-sifting sieve

**Imterqa
taż-żewġ xfafar**
a hatchet with two blades on either side, used for cutting stone slabs into shape

Mannarett
a hatchet

**Imterqa
ta' l-inkolmar**
a hatchet for filling space between one stone and another

Fies tal-ponta
a pickaxe, with one thin-pointed end

Mazza
a mason's iron hammer; this particular one weighs 14 kgs

Invell
a spirit-level

Skwerra falza
a set square used to mark angles other than 90°

Skwerra, skorra
a mason's set square

The Limestone Heritage walk-through

The walk-through provides an understanding as to how stone was quarried and fashioned before delivery to the building site. This walk-through is aided by multi-lingual audio sets which explains the exhibits along the route.

1 Audio-visual presentation

2 Prehistoric times

3 Various quarrying implements, for hundreds of years these tools retained the same shape and use

4 Stone quarring was a hard labour intensive work

5 Separation of stone from the bedrock

6 Cutting of four parallel sections into the long trench in order to form blocks

7 The mule-drawn cart or *karrettun tal-ġebel*

8 Introduction of the mechanized saw

9 Delivery transport by truck made loading easier as well

10 The bell-shaped well (*bir-qanpiena*)

11 Electrically-operated belts increased the efficiency in the loading of slabs

12 The marks left by rock-cutting machines.

13 The remains of a dried-up underwater stream

14 Disused quarries were usually turned into orchards

15 The *girna* is the typical Maltese field hut

16 Typical rural Maltese building features as shown on pages 26, 27, and 28

17 This Limestone Heritage museum houses original tools, fossils, and other material used in the quarrying and building trade.

18 Hands-on experience

19 The walk-through ends in the visitors' centre, where visitors have the opportunity to shop and buy original and exclusive gifts in stone as well as savour local snacks, coffee, and other refreshments.